Annals of the ICRP

ICRP PUBLICATION 105

Radiological Protection in Medicine

Editor
J. VALENTIN

PUBLISHED FOR

The International Commission on Radiological Protection

by

ELSEVIER

Subscription Information

Publication information: *Annals of the ICRP* (ISSN 0146-6453) *Annals of the ICRP*, and the Publications series preceding it, have been published by Pergamon since 1958. In 1991, Pergamon became part of the Elsevier family of imprints. Elsevier carries forward the name and reputation of a publishing house dating from 1580. Elsevier is an integral partner with the scientific, technical and health communities, delivering superior information products and services that foster communication, build insights, and enable individual and collective advancement in scientific research and health care.
Subscription prices are available upon request from the Publisher or from the Regional Sales Office nearest you or from this journal's website (http://intl.elsevierhealth.com/journals/icrp). Further information is available on this journal and other Elsevier products through Elsevier's website: (http://www.elsevier.com). Subscriptions are accepted on a prepaid basis only and are entered on a calendar year basis. Issues are sent by standard mail (surface within Europe, air delivery outside Europe). Priority rates are available upon request. Claims for missing issues should be made within six months of the date of dispatch.

Orders, claims, and journal enquiries: please contact the Regional Sales Office nearest you:

Orlando: Elsevier, Customer Service Department, 6277 Sea Harbor Drive, Orlando, FL 32887-4800, USA; phone: (800) 6542452 [toll free within the USA]; fax: (407) 3639661; phone: (+1) (407) 5636020 [outside the USA]; fax: (+1) (407) 3631354; e-mail: JournalsCustomerService-usa@elsevier.com

Amsterdam: Elsevier, Customer Service Department, PO Box 211, 1000 AE Amsterdam, The Netherlands; phone: (+31) (20) 4853757; fax: (+31) (20) 4853432; e-mail: JournalsCustomerServiceEMEA@elsevier.com

Tokyo: Elsevier, Customer Service Department, 4F Higashi-Azabu, 1-Chome Bldg, 1-9-15 Higashi-Azabu, Minato-ku, Tokyo 106-0044, Japan; phone: (+81) (3) 5561 5037; fax: (+81) (3) 5561 5047; e-mail: JournalsCustomerServiceJapan@elsevier.com

Singapore: Elsevier, Customer Service Department, 3 Killiney Road, #08-01 Winsland House I, Singapore 239519; phone: (+65) 63490222; fax: (+65) 67331510; e-mail: JournalsCustomerServiceAPAC@elsevier.com

Back Issues

This journal is indexed/abstracted in App. Health Phy. Abstr., Biosis Data., CBS and SSSA/CISA/ECA/ISMEC, and Scopus.

∞ ™ The paper used in this publication meets the minimum requirements of American National Standard for Information Sciences – Permanence of Paper for Printed Library Materials, ANSI Z39.48-1984.

USA mailing notice: *Annals of the ICRP* (ISSN 0146-6453) is published bi-monthly (February, April, June, August, October, December) by Elsevier Ltd, The Boulevard, Langford Lane, Kidlington, Oxford OX5 1GB, UK. Periodical postage paid at Rahway NJ and additional mailing offices.

USA POSTMASTER: Send change of address: *Annals of the ICRP*, Elsevier, 6277 Sea Harbor Drive, Orlando, FL 32887-4800.

AIRFREIGHT AND MAILING in USA by Mercury International Limited, 365 Blair Road, Avenel, NJ 07001.

ISBN 978-0-7020-3102-1
ISSN 0146-6453

Published quarterly (December issue)

Printed by Polestar Wheatons Ltd, Exeter, UK

ELSEVIER

ICRP Publication 105

Annals of the ICRP

Radiation Protection in Medicine

ICRP Publication 105

Approved by the Commission in October 2007

Abstract—This report was prepared to underpin the Commission's 2007 Recommendations with regard to the medical exposure of patients, including their comforters and carers, and volunteers in biomedical research. It addresses the proper application of the fundamental principles (justification, optimisation of protection, and application of dose limits) of the Commission's 2007 Recommendations to these individuals.

With regard to medical exposure of patients, it is not appropriate to apply dose limits or dose constraints, because such limits would often do more harm than good. Often, there are concurrent chronic, severe, or even life-threatening medical conditions that are more critical than the radiation exposure. The emphasis is then on justification of the medical procedures and on the optimisation of radiological protection. In diagnostic and interventional procedures, justification of procedures (for a defined purpose and for an individual patient), and management of the patient dose commensurate with the medical task, are the appropriate mechanisms to avoid unnecessary or unproductive radiation exposure. Equipment features that facilitate patient dose management, and diagnostic reference levels derived at the appropriate national, regional, or local level, are likely to be the most effective approaches. In radiation therapy, the avoidance of accidents is a predominant issue. With regard to comforters and carers, and volunteers in biomedical research, dose constraints are appropriate.

Over the last decade, the Commission has published a number of documents that provided detailed advice related to radiological protection and safety in the medical applications of ionising radiation. Each of the publications addressed a specific topic defined by the type of radiation source and the medical discipline in which the source is applied, and was written with the intent of communicating directly with the relevant medical practitioners and supporting medical staff. This report consolidates that advice.

Keywords: Radiological protection; Justification; Optimisation; Patient dose; Dose management

ELSEVIER

ICRP Publication 105

Annals of the ICRP

Editorial

FOUNDATIONS AND FUNDAMENTALS

This space is used, of course, to comment on each new report at the time of its release, and at the same time it provides an opportunity to highlight topical issues and news from the Commission. After some words on the present report, I will come back to some aspects of the Commission's views on tritium below.

Medical exposure IS *unique*. This report is one of the 'Foundation Documents' underpinning the Commission's 2007 Recommendations (ICRP, 2007). However, while Foundation Documents comprising detailed explanations of the biological and physical considerations underlying the Recommendations were published as Annexes A and B of the actual Recommendations, we thought that this summary would fare better as a stand-alone document, particularly for those readers who are directly concerned with medical uses of radiation and want detailed information about medical exposure protection policy.

The main message in the present report is that medical exposure of patients has unique considerations that affect how the fundamental principles are applied. Dose limits are not at all relevant, since ionising radiation, used at the appropriate level of dose for the particular medical purpose, is an essential tool that will cause more good than harm.

Justification in radiological protection of patients is different from justification of other radiation applications, in that generally the very same person enjoys the benefits and suffers the risks associated with a procedure. (There may be other considerations: attendant occupational exposures could be correlated with patient doses or sometimes there can be a trade-off; screening programmes may benefit the population rather than every screened person. But usually, risks and benefits accrue to the same person). And, a very important aspect in daily medical practice: the fact that a method or procedure can be regarded as justified as such does not necessarily mean that its application to the particular patient being considered is justified.

Optimisation of protection for patients is also unique. In the first place, radiation therapy is entirely different from anything else in that the dose to a human being is intentional and its potentially cell-killing properties the very purpose of the treatment. In such cases, optimisation becomes an exercise in minimising doses (and/or their deleterious effects) to surrounding tissues without compromising the pre-determined and intentionally lethal dose and effect to the target volume.

In optimisation of protection of the patient in diagnostic procedures, again the same person gets the benefit and suffers the risk, and again individual restrictions on patient dose could be counterproductive to the medical purpose of the procedure. Therefore, source-related individual dose constraints are not relevant. Instead, Diagnostic Reference Levels (DRLs) for a particular procedure, which apply to groups of similar patients rather than individuals, are used to ensure that doses do not deviate significantly from those achieved at peer departments for that procedure unless there is a known, relevant, and acceptable reason for the deviation. This is in contrast to the Commission's usual balancing of utilitarian protection policies based on collective doses against deontological safeguards using dose constraints for the individual. The policy for radiological protection in medicine is that the radiation exposure be commensurate with the medical purpose.

These various considerations are discussed in some detail in the present report, and in addition, the report includes an overview of the advice provided in the Commission's recent series of topical reports on current issues in medical radiological protection.

The current w_R *for tritium* IS *sufficient*. It is implicit in the above that calculation and application of protection quantities and radiation and tissue weighting factors need careful thought in the medical context. They are, of course, fundamental considerations in all exposure situations, and sometimes the source of heated debate. Recently the relative biological effectiveness (RBE) of beta radiation from tritium and the radiation weighting factor (w_R) to be applied for tritium in routine radiological protection have been the subject of much discussion. To cut a long story short, there is some evidence that at least under some circumstances, an RBE value of 2 for tritium, relative to gamma rays, might be appropriate for cancer induction at low doses, and this in turn has prompted questions as to why ICRP continues to recommend a w_R of 1 for tritium.

Basically, the answer is that for planned exposures, appropriate levels of protection are determined by constrained optimisation, resulting in doses that will be typically a small fraction of the relevant dose limit. There are many uncertainties in the estimation of, e.g., the RBE of tritium radiation, and many intentional simplifications in the assumptions underpinning effective doses which are defined for reference persons and evaluated using reference phantoms. Increased complexity in the calculation of equivalent and effective dose would not improve protection and might suggest a degree of precision in the calculations that is unwarranted.

For a more thorough understanding of the topic, a review of the methodology used in the estimation of doses and risks from internal emitters is provided by Harrison and Day (2008) in the Journal of Radiological Protection. The Commission's position is explained in detail in an invited editorial by Cox (Commission Vice-Chair), Menzel (Committee 2 Chair), and Preston (Committee 1 Chair) in the same issue of that journal. These papers, both of which can be downloaded cost-free from www.iop.org/EJ/journal/JRP, constitute highly recommended reading for those concerned with radiological protection policy issues in relation to internal emitters, particularly tritium.

JACK VALENTIN

References

Cox, R., Menzel, H.-G., Preston, J., 2008. Internal dosimetry and tritium – the ICRP position. J. Radiol. Prot. 28, 131–135.

Harrison, J.D., Day, P., 2008. Radiation doses and risks from internal emitters. J. Radiol. Prot. 28, 137–159.

ICRP, 2007. The 2007 Recommendations of the International Commission on Radiological Protection. ICRP Publication 103. Ann. ICRP 37 (2–4).

CONTENTS

PREFACE

Over the years, the International Commission on Radiological Protection (ICRP), hereinafter referred to as 'the Commission', has issued many reports providing advice on radiological protection and safety in medicine. *Publication 73* was a general overview of this area.

In recent years, the Commission has also addressed some specific situations where difficulties have been observed, trying to produce such topical reports in a style that is accessible to those who may be directly concerned in their daily work, and taking every effort to ensure wide circulation of such reports.

When the 2007 Recommendations were outlined, the Commission concluded that a new general overview report on radiological protection and safety in medicine would be warranted in support of the Recommendations and as an update and amendment to *Publication 73*.

For this particular purpose, the Commission did not launch its usual type of Task Group. Instead, Committee 3 acted as a Task Group in its entirety, but with no outside members. A core group, corresponding to the Full Members of conventional Task Groups, comprised:

C. Cousins	M. Rosenstein (Task Group Chair)	E. Vañó

The other Committee 3 Members, who acted as Corresponding Task Group Members during the preparation of the report, were:

J.-M. Cosset	I. Gusev	Y. Li
J. Liniecki	P. Ortiz López	S. Mattsson
L.V. Pinillos-Ashton	M.M. Rehani	H. Ringertz
C. Sharp (–2006)	Y. Yonekura	

C. Cousins was the Chair of Committee 3, J.-M. Cosset was the Vice-Chair, and E. Vañó was the Committee Secretary.

The report was approved for publication through postal ballot by the Commission in October 2007.

1. BACKGROUND

(1) *Publication 73* (ICRP, 1996), entitled 'Radiological protection and safety in medicine', was published to expand on the application in medicine of the 1990 Recommendations of the Commission (ICRP, 1991a). The current document was prepared by ICRP Committee 3 to augment *Publication 73* and underpin the Commission's 2007 Recommendations (ICRP, 2007d) with regard to the medical exposure of patients, including their comforters and carers, and volunteers in biomedical research.

(2) Over the last decade, the Commission has published a number of documents, prepared by Committee 3, that provide detailed advice related to radiological protection and safety in the medical applications of ionising radiation. Each of these publications addresses a specific topic defined by the type of radiation source and the medical discipline in which the source is applied, and was written with the intent of communicating directly with the relevant medical practitioners and other clinical staff. These publications (in chronological order) are:

- *Publication 84.* Pregnancy and medical radiation (ICRP, 2000a).
- *Publication 85.* Avoidance of radiation injuries from medical intervention procedures (ICRP, 2000b).
- *Publication 86.* Prevention of accidental exposures to patients undergoing radiation therapy (ICRP, 2000c).
- *Publication 87.* Managing patient dose in computed tomography (ICRP, 2000d).
- *Supporting Guidance 2*. Radiation and your patient: a guide for medical practitioners (ICRP, 2001).
- *Supporting Guidance 2*. Diagnostic reference levels in medical imaging – review and additional advice (ICRP, 2001).
- *Publication 93.* Managing patient dose in digital radiology (ICRP, 2003a).
- *Publication 94.* Release of patients after therapy with unsealed radionuclides (ICRP, 2004).
- *Publication 97.* Prevention of high-dose-rate brachytherapy accidents (ICRP, 2005a).
- *Publication 98.* Radiation safety aspects of brachytherapy for prostate cancer using permanently implanted sources (ICRP, 2005b).
- *Publication 102.* Managing patient dose in multi-detector computed tomography (ICRP, 2007c).

(3) Also, in 1999, the Commission published *Publication 80*, entitled 'Radiation dose to patients from radiopharmaceuticals' (ICRP, 1999b); a joint document of Committees 2 and 3 that presented biokinetic and dosimetric data on 10 new radiopharmaceuticals not published previously, and updated the similar data presented in the series of earlier ICRP publications on this subject.

(4) In preparation of the present document, Committee 3:

- reviewed the main topics covered in *Publication 73*;
- augmented that review with the additional advice provided in the documents (listed above) published since *Publication 73*; and
- consulted the draft of the Commission's 2007 Recommendations.

(5) The Commission uses Task Groups and Working Parties to deal with specific areas. Task Groups are appointed by the Commission to perform a defined task, and usually contain a majority of specialists from outside the Commission's structure. Working Parties are set up by Committees with the approval of the Commission to develop ideas for the Committee, sometimes leading to a Task Group. The membership is usually limited to Committee Members. Currently, Committee 3 has a number of similar documents in preparation addressing the following topics:

- radiological protection for cardiologists performing fluoroscopically guided procedures (Task Group);
- evaluation and management of secondary cancer risk in radiation therapy (Joint Task Group with International Commission on Radiation Units and Measurements);
- radiation dose to patients from radiopharmaceuticals (Joint Task Group with Committee 2);
- protecting children: diagnostic techniques involving ionising radiation (Working Party);
- doses to the hands of radiopharmacists (Working Party);
- radiological protection training for diagnostic and fluoroscopically guided interventional procedures (Working Party);
- medical examinations and follow-up of persons accidentally or occupationally exposed to ionising radiation (Working Party); and
- medical screening of asymptomatic persons using ionising radiation (Working Party).

(6) Additional advice from ICRP Committee 3 concerning radiological protection in medicine will be forthcoming as these documents are completed.

(7) In the current report, the term 'exposure' is used to express the act of being exposed to ionising radiation. The terms 'dose' or 'radiation dose' are used when the context is not specific to a particular radiation dose quantity. When the context is specific, the name of the quantity is used (e.g. absorbed dose, equivalent dose, effective dose).

1.1. References

ICRP, 1991a. 1990 Recommendations of the International Commission on Radiological Protection. ICRP Publication 60. Ann. ICRP 21(1–3).

ICRP, 1996. Radiological protection and safety in medicine. ICRP Publication 73. Ann. ICRP 26(2).

ICRP, 1999b. Radiation dose to patients from radiopharmaceuticals. Addendum to ICRP Publication 53. Also includes Addendum 1 to ICRP Publication 72. ICRP Publication 80. Ann. ICRP 28(3).

ICRP, 2000a. Pregnancy and medical radiation. ICRP Publication 84. Ann. ICRP 30(1).

ICRP, 2000b. Avoidance of radiation injuries from medical interventional procedures. ICRP Publication 85. Ann. ICRP 30(2).

ICRP, 2000c. Prevention of accidental exposures to patients undergoing radiation therapy. ICRP Publication 86. Ann. ICRP 30(3).

ICRP, 2000d. Managing patient dose in computed tomography. ICRP Publication 87. Ann. ICRP 30(4).

ICRP, 2001. Radiation and your patient: a guide for medical practitioners. Also includes: Diagnostic reference levels in medical imaging – review and additional advice. ICRP Supporting Guidance 2. Ann. ICRP 31(4).

ICRP, 2003a. Managing patient dose in digital radiology. ICRP Publication 93. Ann. ICRP 34(1).

ICRP, 2004. Release of patients after therapy with unsealed radionuclides. ICRP Publication 94. Ann. ICRP 34(2).

ICRP, 2005a. Prevention of high-dose-rate brachytherapy accidents. ICRP Publication 97. Ann. ICRP 35(2).

ICRP, 2005b. Radiation safety aspects of brachytherapy for prostate cancer using permanently implanted sources. ICRP Publication 98. Ann. ICRP 35(3).

ICRP, 2007c. Managing patient dose in multi-detector computed tomography. ICRP Publication 102. Ann. ICRP 37(1).

ICRP, 2007d. The 2007 Recommendations of the International Commission on Radiological Protection. ICRP Publication 103. Ann. ICRP 37(2–4).

2. USE OF IONISING RADIATION IN MEDICINE

(8) More people are exposed to ionising radiation from medical practice than from any other human activity, and in many cases, the individual doses are higher. In countries with advanced healthcare systems, the annual number of radiological diagnostic procedures approaches or exceeds 1 for every member of the population (UNSCEAR, 2000). Furthermore, doses to patients for the same type of examination differ widely between centres, suggesting that there is considerable scope for management of patient dose (UNSCEAR, 2000).

(9) Radiation exposures in medicine are predominantly to individuals undergoing diagnostic examinations, interventional procedures, or radiation therapy. Diagnostic examinations include those for medical and dental purposes. Interventional procedures are predominantly fluoroscopically guided, but computed tomography guided techniques are also being developed and utilised. However, staff, and other individuals helping to support and comfort patients, are also exposed to radiation. The other individuals include parents holding children during diagnostic procedures, and family or close friends who may come close to patients following the administration of radiopharmaceuticals or during brachytherapy. Exposure to members of the general public resulting from the use of radiation in medicine also occurs, but it is almost always at very low levels. Other Commission documents cover radiological protection for workers in medicine (occupational exposure), and radiological protection for members of the general public associated with medicine (public exposure), but some brief comments on these topics are given in Sections 16.1 and 16.2. The rest of this document concentrates on medical exposure of patients, their comforters and carers, and volunteers in biomedical research, as described below.

- The exposure of individuals for diagnostic, interventional, and therapeutic purposes, including exposure of the embryo/fetus or infant during medical exposure of patients who are pregnant or breastfeeding.
- Exposures (other than occupational) incurred knowingly and willingly by individuals, such as family and close friends (or other comforters), helping either in hospital or at home in the support and comfort of patients undergoing diagnosis or treatment.
- Exposures incurred by volunteers as part of a programme of biomedical research that provides no direct benefit to the volunteers.

(10) The use of radiation for medical exposure of patients contributes over 95% of man-made radiation exposure and is only exceeded world-wide by natural background as a source of exposure (UNSCEAR, 2000). In a preliminary analysis for 2006 in the United States, the contribution of medical exposure of patients is expected to be similar in magnitude to natural background as a source of exposure to the U.S. population (Mettler et al., 2008).

(11) UNSCEAR (2000) compared estimates of the 1985–1990 and 1991–1996 periods, and concluded that the worldwide annual per caput effective dose from medical exposure of patients increased by 35% and the collective dose increased by 50%, while the population increased by only 10%. It was also estimated that, worldwide,

there were approximately 2000 million x-ray studies, 32 million nuclear medicine studies, and over 6 million radiation therapy patients treated annually. These numbers are expected to increase in future years.

(12) Overall, medical exposure has increased since the UNSCEAR (2000) evaluation, largely due to the rapid increase in the utilisation of computed tomography (CT), both in industrialised and in developing countries (ICRP, 2000d; ICRP, 2007c).

(13) Worldwide, the estimated number of medical and dental radiographic machines is approximately 2 million. While it is difficult to estimate the number of occupationally exposed medical workers, UNSCEAR (2000) estimated that there are more than 2.3 million monitored medical radiation workers.

2.1. References

ICRP, 2000d. Managing patient dose in computed tomography. ICRP Publication 87. Ann. ICRP 30(4).

ICRP, 2007c. Managing patient dose in multi-detector computed tomography. ICRP Publication 102. Ann. ICRP 37(1).

Mettler, F.A., Thomadsen, B.R., Bhargavan, M., et al., 2008. Medical radiation exposure in the U.S. 2006: Preliminary results. 43rd Annual Meeting of the National Council on Radiation Protection and Measurements: Advances in Radiation Protection in Medicine. Health Phys. 95, in press.

UNSCEAR, 2000. Sources and Effects of Ionising Radiation. United Nations Scientific Committee on the Effects of Atomic Radiation Report to the General Assembly with Scientific Annexes. United Nations, New York, NY.

3. BRIEF SUMMARY OF BIOLOGICAL BASIS FOR RADIOLOGICAL PROTECTION

(14) The biological effects of radiation can be grouped into two types: deterministic effects (tissue reactions) and stochastic effects (cancer and heritable effects). These effects are noted briefly here; the biological basis for radiological protection is covered in depth in the 2007 Recommendations and other Commission documents.

3.1. Deterministic effects (tissue reactions)

(15) If the effect only results when many cells in an organ or tissue are killed, the effect will only be clinically observable if the radiation dose is above some threshold. The magnitude of this threshold will depend on the dose rate (i.e. dose per unit time) and linear energy transfer of the radiation, the organ or tissue irradiated, the volume of the irradiated part of the organ or tissue, and the clinical effect of interest. With increasing doses above the threshold, the probability of occurrence will rise steeply to 100% (i.e. every exposed person will show the effect), and the severity of the effect will increase with dose. The Commission calls these effects 'deterministic' (tissue reactions), and a detailed discussion and information on deterministic effects (tissue reactions) is found in ICRP (2007a). Such effects can occur in the application of ionising radiation in radiation therapy, and in interventional procedures, particularly when fluoroscopically guided interventional procedures are complex and require longer fluoroscopy times or acquisition of numerous images.

3.2. Stochastic effects (cancer and heritable effects)

(16) There is good evidence from cellular and molecular biology that radiation damage to the DNA in a single cell can lead to a transformed cell that is still capable of reproduction. Despite the body's defences, which are normally very effective, there is a small probability that this type of damage, promoted by the influence of other agents not necessarily associated with radiation, can lead to a malignant condition (somatic effect). As the probability is low, this will only occur in a few of those exposed. If the initial damage is to the germ cells in the gonads, heritable effects may occur.

(17) The probability of a stochastic effect attributable to the radiation increases with dose and is probably proportional to dose at low doses. At higher doses and dose rates, the probability often increases with dose more markedly than simple proportion. At even higher doses, close to the thresholds of deterministic effects (tissue reactions), the probability increases more slowly, and may begin to decrease, because of the competing effect of cell killing. These effects, both somatic and heritable, are called 'stochastic'. The probability of such effects is increased when ionising radiation is used in medical procedures.

(18) Although a single radiological examination only leads to a small increase in the probability of cancer induction in a patient, in industrialised countries each

member of the population undergoes, on average, one such examination each year; therefore, the cumulative risk increases accordingly. Calculations performed on the assumption of a linear non-threshold model of radiation action estimate that the proportion of cancer deaths in a general population that could be attributed to exposure from radiological procedures may reach a level from a fraction of one to a few percent of that cancer mortality (NAS/NRC, 2006). In addition, the risk is non-uniformly distributed in a population. Some groups of patients are examined much more frequently due to their health status. Also, some groups show higher than average sensitivity for cancer induction (e.g. embryo/fetus, infants, young children, those with genetic susceptibility). Moreover, cancers occurring early in life result in much higher lifetime loss than cancers that become manifest late in life. All these circumstances indicate that proper justification of radiation use and optimisation of radiation protection in medicine are indispensable principles of radiological protection.

(19) A detailed discussion and information on somatic and heritable effects is found in ICRP (2007a), and the Commission's view on cancer risk at low doses is presented in *Publication 99* (ICRP, 2005c). It is not feasible to determine on epidemiological grounds alone that there is, or is not, an increased risk of cancer for members of the public associated with absorbed doses of the order of 100 mGy or below. The linear non-threshold model remains a prudent basis for the practical purposes of radiological protection at low doses and low dose rates.

(20) The Commission has also reviewed the topic of individuals with genetic susceptibility to cancer, and expressed its preliminary view in *Publication 79* (ICRP, 1999a) that the information available is insufficient to provide a meaningful quantitative judgement on this issue. The Commission will continue to monitor this subject with regard to its implications for radiological protection.

3.3. Effects of in-utero irradiation

(21) There are radiation-related risks to the embryo/fetus during pregnancy that are related to the stage of pregnancy and the absorbed dose to the embryo/fetus. These are noted below briefly under the topics of lethal effects, malformations, central nervous system effects, and leukaemia and childhood cancer. The Commission has evaluated the effects of prenatal irradiation in detail in *Publication 90* (ICRP, 2003b).

3.3.1. Lethal effects

(22) There is embryonic sensitivity to the lethal effects of irradiation in the preimplantation period of embryonic development. At doses below 100 mGy, such lethal effects will be very infrequent and there is no reason to believe that significant risks to health will express after birth.

3.3.2. Malformations

(23) During the period of major organogenesis, conventionally taken to be from the third to the eighth week after conception, malformations may be caused, particularly in the organs under development at the time of exposure. These effects have a threshold of approximately 100 mGy.

3.3.3. Central nervous system

(24) From 8 to 25 weeks after conception, the central nervous system is particularly sensitive to radiation. A reduction in intelligence quotient cannot be identified clinically at fetal doses below 100 mGy. During the same time period, fetal doses in the range of 1 Gy result in a high probability of severe mental retardation. The sensitivity is highest from 8 to 15 weeks after conception, and lower from 16 to 25 weeks of gestational age.

3.3.4. Leukaemia and childhood cancer

(25) Radiation has been shown to increase the probability of leukaemia and many types of cancer in both adults and children. Throughout most of pregnancy, the embryo/fetus is assumed to be at approximately the same risk for potential carcinogenic effects as children (i.e., about three times that of the population as a whole).

(26) Consideration of the effects listed above is important when pregnant patients undergo diagnostic examinations, interventional procedures, and radiation therapy using ionising radiation. A balance must be attained between the health care of the patient and the potential for detrimental health effects to the embryo/fetus that accompanies the specific radiological procedure.

3.4. References

ICRP, 1999a. Genetic susceptibility to cancer. ICRP Publication 79. Ann. ICRP 28(1/2).

ICRP, 2003b. Biological effects after prenatal irradiation (embryo and fetus). ICRP Publication 90. Ann. ICRP 33(1/2).

ICRP, 2005c. Low-dose extrapolation of radiation-related cancer risk. ICRP Publication 99. Ann. ICRP 35(4).

ICRP, 2007a. Biological and epidemiological information on health risks attributable to ionising radiation: a summary of judgements for the purposes of radiological protection of humans. Annex A to 2007 Recommendations.

NAS/NRC, 2006. Health Risks from Exposure to Low Levels of Ionising Radiation: BEIR VII Phase 2. Board on Radiation Effects Research. National Research Council of the National Academies, Washington, D.C.

4. DOSIMETRIC QUANTITIES

(27) The basic physical quantity used in radiological protection for stochastic effects is the absorbed dose averaged over an organ or tissue (i.e. mean absorbed dose; the energy deposited in the organ divided by the mass of that organ or tissue). For deterministic effects (tissue reactions), the absorbed dose is averaged over the highly irradiated portion of the tissue, such as the volume of irradiated skin in the direct radiation field. The SI unit for absorbed dose is joule per kilogramme (J/kg) and its special name is 'gray' (Gy).

(28) During medical imaging procedures using x rays, absorbed doses in organs or tissues of the patient undergoing diagnostic or interventional procedures cannot usually be measured directly. Therefore, measurable quantities that characterise the external radiation field are used to assist in managing the patient dose. These include simple quantities such as absorbed dose in a tissue-equivalent material at the surface of a body or in a phantom, but also a number of other quantities of varying complexity, depending on the nature of the x-ray equipment [e.g. for CT, see ICRP (2000d, 2007c)]. Significant progress has been achieved in recent years in providing methods to derive absorbed doses in tissues and organs from a number of practical measurements, and a considerable body of data is available, in particular that found in ICRU Report 74, 'Patient dosimetry for x rays used in medical imaging' (ICRU, 2005). In nuclear medicine, administered activity [in becquerels (Bq)] is the measurable quantity used.

(29) Some radiations are more effective than others in causing stochastic effects. To allow for this, a quantity equivalent dose (the average absorbed dose in an organ or tissue multiplied by a dimensionless radiation weighting factor) has been introduced. For all the principal radiations used in medicine (photons and electrons), the radiation weighting factor is assigned a value of 1, so the absorbed dose and the equivalent dose are numerically equal. For alpha particles and heavy ions, the radiation weighting factor is 20, for protons, the radiation weighting factor is 2, and for neutrons, the radiation weighting factor is a continuous function of the neutron energy incident on the body. The special name for the unit of equivalent dose is sievert (Sv). A detailed discussion on radiation weighting factors is provided in *Publication 92* (ICRP, 2003c).

(30) Radiation exposure of the different organs and tissues in the body results in different probabilities of harm and different severities. The Commission calls the combination of probability and severity of harm 'detriment', meaning health detriment. To reflect the combined detriment from stochastic effects due to the equivalent doses in all the organs and tissues of the body, the equivalent dose in each organ and tissue is multiplied by a tissue weighting factor, and the results are summed over the whole body to give the effective dose. The special name for the unit of effective dose is also sievert (Sv). The tissue weighting factors in the 2007 Recommendations are those recommended in ICRP (2007b).

(31) The Commission intended effective dose for use as a principal protection quantity for the establishment of radiological protection guidance. It should not be used to assess risks of stochastic effects in retrospective situations for exposures

in identified individuals, nor should it be used in epidemiological evaluations of human exposure, because the Commission has made judgements on the relative severity of various components of the radiation risks in the derivation of 'detriment' for the purpose of defining tissue weighting factors. Such risks for stochastic effects are dependent on age and sex. The age and sex distributions of workers and the general population (for which the effective dose is derived) can be quite different from the overall age and sex distribution for the population undergoing medical procedures using ionising radiation, and will also differ from one type of medical procedure to another, depending on the prevalence of the individuals for the medical condition being evaluated. For these reasons, risk assessment for medical uses of ionising radiation is best evaluated using appropriate risk values for the individual tissues at risk, and for the age and sex distribution of the individuals undergoing the medical procedures.

(32) Effective dose can be of practical value for comparing the relative doses related to stochastic effects from:

- different diagnostic examinations and interventional procedures;
- the use of similar technologies and procedures in different hospitals and countries; and
- the use of different technologies for the same medical examination;

provided that the representative patients or patient populations for which the effective doses are derived are similar with regard to age and sex. However, comparisons of effective doses derived as given in Section 4.3.5 of the Commission's 2007 Recommendations (ICRP, 2007d) are inappropriate when there are significant dissimilarities between the age and sex distributions of the representative patients or patient populations being compared (e.g., children, all females, elderly patients) and the Commission's reference distribution of both sexes and all ages. This is a consequence of the fact that the magnitudes of risk for stochastic effects are dependent on age and sex.

4.1. References

ICRP, 2000d. Managing patient dose in computed tomography. ICRP Publication 87. Ann. ICRP 30(4).

ICRP, 2003c. Relative biological effectiveness (RBE), quality factor (Q), and radiation weighting factor (w_R). ICRP Publication 92. Ann. ICRP 33(4).

ICRP, 2007b. Quantities used in radiological protection. Annex B to 2007 Recommendations.

ICRP, 2007c. Managing patient dose in multi-detector computed tomography. ICRP Publication 102. Ann. ICRP 37(1).

ICRP, 2007d. The 2007 Recommendations of the International Commission on Radiological Protection. ICRP Publication 103. Ann. ICRP 37(2–4).

ICRU, 2005. Patient dosimetry for x rays used in medical imaging. ICRU Report 74. J. ICRU 5(2).

5. FRAMEWORK OF RADIOLOGICAL PROTECTION IN THE 2007 RECOMMENDATIONS

(33) The primary aim of radiological protection is to provide an appropriate standard of protection for people and the environment without unduly limiting the beneficial practices giving rise to radiation exposure. As noted before, medical radiation sources are used deliberately in the health care of patients and are designed for use in a controlled manner.

(34) In its 2007 Recommendations (ICRP, 2007d), the Commission has formulated a set of principles that apply equally to planned, emergency, and existing situations, and clarifies how the fundamental principles (justification, optimisation of protection, and application of dose limits) apply to radiation sources and to the individual, as well as how the source-related principles (justification and optimisation of protection) apply to all exposure situations.

5.1. Source-related principles

(35) The following two source-related principles apply in all exposure situations.

- The principle of justification: any decision that alters the existing radiation exposure situation (e.g. by introducing a new radiation source or by reducing existing exposure) should do more good than harm. This means that by introducing a new radiation source, by reducing existing exposure, or by reducing the risk of potential exposure, one should achieve sufficient individual or societal benefit to offset the detriment it causes.
- The principle of optimisation of protection: the likelihood of incurring exposures, the number of people exposed, and the magnitude of their individual doses should all be kept as low as reasonably achievable, taking into account economic and societal factors. This means that the level of protection should be the best under the prevailing circumstances, maximising the margin of benefit over harm. In order to avoid severely inequitable outcomes of this optimisation procedure, there should be restrictions on the doses or risks to the individuals from a particular source (dose or risk constraints and reference levels).

The Commission uses 'dose constraint' in planned exposure situations and 'reference level' for existing and emergency exposure situations. However, although the medical exposure of patients is a planned situation, the dose constraint is not applicable and the diagnostic reference level (Section 10) is used as a tool for the optimisation of protection in medical exposure of patients.

5.2. Individual-related principle

(36) This principle applies in planned exposure situations, except medical exposure of patients.

- The principle of application of dose limits in planned situations: the total dose to any individual from all the regulated sources in planned situations other than medical exposure of patients should not exceed the appropriate limits recommended by the Commission.

(37) Provided that the medical exposures of patients have been properly justified and that the associated doses are commensurate with the medical purpose, it is not appropriate to apply dose limits or dose constraints to the medical exposure of patients, because such limits or constraints would often do more harm than good (see Sections 9.2 and 11).

(38) In most situations in medicine, other than radiation therapy, it is not necessary to approach the thresholds for deterministic effects (tissue reactions), even for the most part in fluoroscopically guided interventional procedures, if the staff are properly educated and trained. The Commission's policy is therefore to limit exposures so as to keep doses below these thresholds. The possibility of stochastic effects cannot be eliminated totally, so the policy is to avoid unnecessary sources of exposure and to take all reasonable steps to reduce the doses from those sources of exposure that are necessary or cannot be avoided.

(39) In using these principles to develop a practical system of radiological protection that fits smoothly into the conduct of the activity, the Commission uses a division into three types of exposure: medical exposure, which is principally the exposure of persons as part of their diagnosis or treatment (or exposure of a patient's embryo/fetus or breast-feeding infant) and their comforters and carers (other than occupational), but also includes volunteers in biomedical research; occupational exposure, which is exposure incurred at work and principally as a result of work; and public exposure, which comprises all other exposures. In some respects, the system of protection is applied differently to these types of exposure, so it is important to clarify the distinctions. The distinctions concerning medical exposure to patients, comforters and carers (other than occupational), and volunteers in biomedical research (as described in Section 2) are covered in this document.

5.3. Reference

ICRP, 2007d. The 2007 Recommendations of the International Commission on Radiological Protection. ICRP Publication 103. Ann. ICRP 37(2–4).

6. UNIQUE ASPECTS OF RADIOLOGICAL PROTECTION IN MEDICINE FOR PATIENTS

(40) Several features of radiation exposure in medicine for patients require an approach to radiological protection that is somewhat different from that for other types of radiation exposure.

6.1. Deliberate exposure

(41) The exposure of patients is deliberate. Except in radiation therapy, it is not the aim to deliver radiation dose, but rather to use the radiation to provide diagnostic information or to conduct an interventional procedure. Nevertheless, the dose is given deliberately and cannot be reduced indefinitely without prejudicing the intended outcome.

6.2. Voluntary exposure

(42) Medical uses of radiation for patients are voluntary in nature, combined with the expectation of direct individual health benefit to the patient. The voluntary decision is made with varying degrees of informed consent that includes not only the expected benefit but also the potential risks (including radiation). The amount of information provided in order to obtain informed consent varies based on the exposure level (e.g. whether diagnostic, interventional, or therapeutic) and the possible emergent medical circumstances that may be attributable to radiation exposure. Generally, little informed consent is obtained for low-risk procedures (such as a chest x-ray procedure), more informed consent is obtained for interventional procedures, and a high level (typically written) of consent is often obtained before most radiation therapy.

(43) The exception to the concept of a voluntary exposure leading to a direct individual medical benefit is the use of radiation in biomedical research. In these circumstances, the voluntary exposure usually accrues to a societal benefit rather than an individual benefit. Informed consent is always needed.

6.3. Medical screening of asymptomatic patients

(44) Screening is performed with the aim of identifying a disease process that has not become manifest clinically. The aim is that earlier diagnosis will lead to earlier and more effective treatment and a better outcome in terms of quality of life and survival. For example, current screening practices using ionising radiation (e.g. mammography) appear to be valid and are recommended for certain populations. On the other hand, there is increasing use of CT (including self-referral) and positron emission tomography in screening for disease in asymptomatic individuals, and most of these screening applications have not been justified on the basis of current scientific literature.

(45) Patients undergoing screening should be fully informed of the potential benefits and risks, including the radiation risks. Each application of ionising radiation for screening of asymptomatic individuals should be evaluated and justified with regard to its clinical merit.

6.4. Radiation therapy

(46) In radiation therapy, the aim is to eradicate the neoplastic target tissue or to palliate the patient's symptoms. Some deterministic damage (tissue reactions) to surrounding tissue and some risk of stochastic effects in exposed non-target tissues are inevitable, but the goal of all radiation therapy is to optimise the relationship between the probability of tumour control and normal tissue complications.

6.5. Management of radiation dose

(47) In medicine, the requirement is to manage the radiation dose to the patient to be commensurate with the medical purpose. The goal is to use the appropriate dose to obtain the desired image or desired therapy. In this regard, the Commission introduced the use of diagnostic reference levels for imaging procedures, which will be discussed in more detail later in this report.

6.6. Demographics of the patient population

(48) Risk estimates developed by the Commission apply to either the working population or the whole population, and were derived for age- and sex-averaged populations for the purpose of establishing radiological protection guidance (see Section 4). The risks for various age groups differ depending on the age at exposure and the organs and tissues exposed. For the exposure of young children, the attributable lifetime risk of death (total cancers) would be higher, perhaps by a factor of 2 or 3 (ICRP, 1991a). For many common types of diagnostic examination, the higher risk per unit dose may be offset by the reduction in dose relative to the dose received by an adult. For an age at exposure of approximately 60 years, the risk would be lower, perhaps by a factor of 3. At higher ages at exposure, the risks are even lower (ICRP, 1991a).

(49) It is difficult to apply the concept of effective dose to compare doses from medical exposure of patients to other sources of exposure to humans, as the effective dose values for the other sources are for an age- and sex-averaged population. Effective dose can be of value for comparing doses from different diagnostic procedures and for comparing the use of similar technologies and procedures in different hospitals and countries, as well as the use of different technologies for the same medical examination, provided that the reference patient or patient populations are similar with regard to age and sex. As noted in Section 4, for planning the exposure of patients and risk–benefit assessments, the equivalent dose or the absorbed dose to irradiated tissues is the relevant quantity.

6.7. Range of detriments from radiation uses in medicine

(50) A wide range of potential radiation detriment to an individual patient occurs in medical practice. The detriments range from (most commonly) minimal to (rarely) lethal.

(51) An example of minimal detriment would be a chest x-ray procedure on a very elderly patient. There would be no chance, due to the low absorbed dose, of deterministic effects (tissue reactions) and, in view of the patient's age, essentially no risk of stochastic effects being expressed in the individual's lifetime.

(52) An example of more significant potential for radiation detriment is CT examinations, which can involve relatively high doses to patients. The absorbed doses to tissues from a whole-body CT examination are typically in the range of 10–100 mGy. Therefore, a non-symptomatic 45-year-old adult who, beginning at that age, undergoes voluntary (self-referred) annual whole-body CT examinations for 30 years could accrue a significant lifetime cumulative absorbed dose to tissues [i.e. 300–3000 mGy (0.3–3 Gy)]. This cumulative absorbed dose is of a magnitude at which an increase in the probability of cancer has been observed in human epidemiological studies (UNSCEAR, 2000).

(53) There is also a growing number of deterministic injuries (tissue reactions) resulting from unnecessarily high doses from the use of fluoroscopy during interventional procedures (ICRP, 2000b). In addition, in radiation oncology, the tolerance for deviation from the treatment regimen is very small. Usually, overdosage in excess of 10% will result in an unacceptably high risk of complications. Underdosage will result in failure to cure the cancer and will lead to more cancer deaths than expected.

6.8. References

ICRP, 1991a. 1990 Recommendations of the International Commission on Radiological Protection. ICRP Publication 60. Ann. ICRP 21(1–3).

ICRP, 2000b. Avoidance of radiation injuries from medical interventional procedures. ICRP Publication 85. Ann. ICRP 30(2).

UNSCEAR, 2000. Sources and Effects of Ionising Radiation. United Nations Scientific Committee on the Effects of Atomic Radiation Report to the General Assembly with Scientific Annexes, United Nations, New York, NY.

7. DISCUSSION OF THE TERM 'PRACTICE'

(54) The Commission previously distinguished between 'practices' that added doses and 'interventions' that reduced doses (ICRP, 1991a). Different protection principles were applied in the two situations. That distinction caused difficulties and was seen as artificial. The Commission now recommends one set of principles for all the situations to which its recommendations apply, namely planned, emergency, and existing situations.

(55) The term 'practice' has, however, become widely used in radiological protection. In the field of medicine, the term 'practice' typically refers to the medical care that a practitioner provides to patients. For example, in radiation oncology, the term refers to initial consultation with the patient, accurate diagnosis and staging of the cancer, treatment planning, administering a course of treatment, and subsequent follow-up.

(56) The term 'practice', when the Commission is communicating with the medical community regarding the utilisation of ionising radiation in medicine, needs to be presented in a way that is readily understood by the medical community. The term 'radiological practice in medicine' is recommended in order to differentiate between the usual meaning of 'practice' in medicine. This should help the medical profession to better understand the radiological protection concepts of the Commission.

7.1. Reference

ICRP, 1991a. 1990 Recommendations of the International Commission on Radiological Protection. ICRP Publication 60. Ann. ICRP 21(1–3).

8. JUSTIFICATION OF A RADIOLOGICAL PRACTICE IN MEDICINE

(57) In principle, the decision to adopt or continue any human activity involves a review of the benefits and disadvantages of the possible options. This review usually provides a number of alternative procedures that will do more good than harm. The more elaborate process of judging which of these options is the 'best' (e.g. choosing between the use of x rays or ultrasound) is still necessary and is more complex. The harm, more strictly the detriment, to be considered is not confined to that associated with the radiation; it includes other detriments and the economic and societal costs of the practice. Often, the radiation detriment will be only a small part of the total. For these reasons, the Commission limits its use of the term 'justification' to the first of the above stages (i.e. it requires only that the net benefit be positive). Searching for the best available option is usually a task beyond the responsibility of radiological protection organisations.

(58) Depending on the healthcare system in a country, there may be an influence of commercial interests on referral of patients to radiological examinations, since such examinations may be a major source of income to hospitals, academic medical institutions, and clinics with modern radiological departments. Such a situation may create referral incentives for frequent radiological examinations of patients that could exceed the needs of good medical practice. The Commission disapproves of such referrals that confer unjustifiable risk on patients, being inconsistent with medical ethics and principles of radiological protection.

(59) Most of the assessments needed for the justification of a radiological practice in medicine are made on the basis of experience, professional judgement, and common sense. However, quantitative decision-aiding techniques are available and, if the necessary data are accessible, they should be considered.

(60) There are three levels of justification of a radiological practice in medicine.

- At the first and most general level, the proper use of radiation in medicine is accepted as doing more good than harm to society. This general level of justification is now taken for granted, and is not discussed here further.
- At the second level, a specified procedure with a specified objective is defined and justified (e.g. chest x rays for patients showing relevant symptoms, or a group of individuals at risk for a condition that can be detected and treated). The aim of the second level of justification is to judge whether the radiological procedure will improve the diagnosis or treatment, or will provide necessary information about the exposed individuals.
- At the third level, the application of the procedure to an individual patient should be justified (i.e. the particular application should be judged to do more good than harm to the individual patient). Hence all individual medical exposures should be justified in advance, taking into account the specific objectives of the exposure and the characteristics of the individual involved.

(61) The second and third levels of justification are discussed below.

8.1. Justification of a defined radiological procedure (Level 2)

(62) The justification of a radiological procedure is a matter for national and international professional bodies, in conjunction with national health and radiological protection authorities, and the corresponding international organisations. The total benefits from a medical procedure include not only the direct health benefits to the patient, but also the benefits to the patient's family and to society.

(63) It should be noted that the justification of a medical procedure does not necessarily lead to the same choice of the best procedure in all situations. For example, chest fluoroscopy for the diagnosis of serious pulmonary conditions may do more good than harm, but chest radiography is likely to be the procedure of choice in a country with substantial resources, because the ratio of good to harm would be larger. However, fluoroscopy may be the procedure chosen in developing countries with fewer resources, if it would still produce a net benefit and if no better alternatives were available.

(64) In a similar manner, the justification for routine radiological screening for some types of cancer will depend on the national incidence and on the availability of effective treatment for detected cases. National variations are to be expected.

(65) Although the main exposures in medicine are to patients, the exposures to staff and to members of the public who are not connected with the procedures should be considered. The possibility of accidental or unintended exposures should also be considered. The decisions should be reviewed from time to time, as more information becomes available about the risks and effectiveness of the existing procedure and about new procedures.

(66) The justification of diagnostic investigations for which the benefit to the patient is not the primary objective needs special consideration. In the use of radiography for insurance purposes, the primary benefit usually accrues to the insurer, but there may be some economic benefit for the individual examined. Examinations ordered by physicians as a defence against malpractice claims may only have marginal advantages for the individual patient.

8.2. Justification of a procedure for an individual patient (Level 3)

(67) Justification of individual exposures should include checking that the required information is not already available. Usually, no additional justification is needed for the application of a simple diagnostic procedure to an individual patient with the symptoms or indications for which the procedure has already been justified in general. For high-dose examinations, such as complex diagnostic and interventional procedures, individual justification by the practitioner is particularly important and should take account of all the available information. This includes the details of the proposed procedure and of alternative procedures, the characteristics of the individual patient, the expected dose to the patient, and the availability of information on previous or expected examinations or treatment. It will often be possible to speed up the procedure by defining referral criteria and patient categories in advance.

9. OPTIMISATION OF PROTECTION FOR PATIENTS IN MEDICAL EXPOSURES

9.1. General approach

(68) The optimisation of radiological protection for patients in medicine is usually applied at two levels: (1) the design, appropriate selection, and construction of equipment and installations; and (2) the day-to-day methods of working (i.e. the working procedures). The basic aim of this optimisation of protection is to adjust the protection measures for a source of radiation in such a way that the net benefit is maximised.

(69) The concepts involved can be set out in simple terms, but their practical application can range from simple common sense to complex quantitative processes. In selecting the provision for protection in relation to a source, there is always a choice of options. The choice of protection option directly alters the level of exposure of the patient, the staff, and sometimes the public. However, the choice also alters the scale of resources applied to protection. These resources may be reflected directly in financial costs, but they may also involve less easily quantified societal costs such as other health risks to staff.

(70) The optimisation of radiological protection means keeping the doses 'as low as reasonably achievable, economic and societal factors being taken into account', and is best described as management of the radiation dose to the patient to be commensurate with the medical purpose.

9.2. Use of diagnostic reference levels and dose constraints

(71) In protection of the patient, the detriments and benefits are received by the same individual, the patient, and the dose to the patient is determined principally by the medical needs. Dose constraints for patients are therefore inappropriate, in contrast to their importance in occupational and public exposure. Nevertheless, management of patient dose is important and can often be facilitated for diagnostic and interventional procedures by use of a diagnostic reference level, which is a method for evaluating whether the patient dose (with regard to stochastic effects) is unusually high or low for a particular medical imaging procedure (Section 10).

(72) In the exposure of comforters and carers (other than occupational), and in the exposure of volunteers in biomedical research programmes that provide no direct benefit to the volunteers, dose constraints are applicable to limit inequity and because there is no further protection in the form of a dose limit.

9.3. Management of medical exposures

(73) There is considerable scope for dose reductions in diagnostic radiology. Simple, low-cost measures are available to reduce doses without loss of diagnostic information, but the extent to which these measures are used varies widely.

(74) The optimisation of protection in medical exposures (as implemented through management of patient dose) does not necessarily mean the reduction of doses to the patient. For example, diagnostic radiographic equipment often uses antiscatter grids to improve the image quality, yet removing the grid would allow a reduction in dose by a factor of 2–4. For radiography of the abdomen of adults, where the scattered radiation is important, the net benefit would be reduced by removing the grid because the benefit of the dose reduction would be more than offset by the loss of quality of the image. The optimisation of protection would not call for the removal of the grid. In the radiography of small children, however, the amount of scattered radiation is less and the benefit of the dose reduction resulting from the removal of the grid is not fully offset by the small deterioration of the image. The optimisation of protection then calls for the reduction in dose allowed by the removal of the grid.

(75) In radiation therapy, it is necessary to differentiate between the dose to the target tissue and the dose to other parts of the body. If the dose to the target tissue is too low, the therapy will be ineffective. The exposures will not have been justified and the optimisation of protection does not arise. However, the protection of tissues outside the target volume is an integral part of dose planning, which can be regarded as including the same aims as the optimisation of protection.

(76) The exposure (other than occupational) of individuals helping to support and comfort patients includes the exposure of families and friends of patients discharged from hospital after therapeutic nuclear medicine procedures using unsealed radionuclides or permanently implanted sealed sources. The procedure of optimisation of protection for these groups is no different from that for public exposure, except that the exposures need not be restricted by dose limits, but include the use of dose constraints.

10. DIAGNOSTIC REFERENCE LEVELS

(77) Guidance for the use of diagnostic reference levels for patients in medical exposure has been provided in *Publication 60* (ICRP, 1991a), *Publication 73* (ICRP, 1996), and *Supporting Guidance 2* (ICRP, 2001). Summaries of that guidance, which include some of the history of the evolution of the concept of diagnostic reference levels, are given in this section.

10.1. Diagnostic reference levels (*Publications 60* and *73*)

(78) In *Publication 60* (ICRP, 1991a), reference levels were described as values of measured quantities above which some specified action or decision should be taken. They include recording levels, above which a result should be recorded, lower values being ignored; investigation levels, above which the cause or the implications of the result should be examined; intervention levels, above which some remedial action should be considered; and, more generally, action levels, above which some specified action should be taken. The use of these levels can avoid unnecessary or unproductive work and can help in the effective deployment of resources. They can also be helpful in radiological protection by drawing attention to situations of potentially high risk.

(79) One particular form of reference level (the diagnostic reference level) applies to medical x-ray imaging and diagnostic nuclear medicine. In *Publication 60* (ICRP, 1991a), the Commission recommended that consideration should be given to the use of dose constraints, or investigation levels, selected by the appropriate professional organisation or regulatory authority, for application in some common diagnostic procedures. They should be applied with flexibility, to allow higher doses where indicated by sound clinical judgement. In *Publication 73* (ICRP, 1996), the Commission decoupled the concept of diagnostic reference level from that of a dose constraint, and discussed the concept in more detail, as noted below.

(80) Diagnostic reference levels, which are a form of investigation level, apply to an easily measured quantity, usually the absorbed dose in air, or in a tissue-equivalent material at the surface of a simple standard phantom or representative patient. In nuclear medicine, the quantity will usually be the administered activity. In both cases, the diagnostic reference level will be intended for use as a simple test for identifying situations where the levels of patient dose or administered activity are unusually high or low.

(81) If it is found that procedures are consistently causing the relevant diagnostic reference level to be exceeded, there should be a local review of the procedures and the equipment in order to determine whether the protection has been adequately optimised. If not, measures aimed at reduction of the doses should be taken.

(82) Diagnostic reference levels are supplements to professional judgement and do not provide a dividing line between 'good' and 'bad' medicine. They contribute to good radiological practice in medicine. The numerical values of diagnostic reference levels are advisory; however, implementation of the diagnostic reference level concept may be required by an authorised body (ICRP, 2001). It is inappropriate

to use the numerical values for diagnostic reference levels as regulatory limits or for commercial purposes.

(83) Diagnostic reference levels apply to radiation exposure of patients resulting from medical x-ray imaging and diagnostic nuclear medicine procedures. They do not apply to radiation therapy. Diagnostic reference levels have no direct linkage to the numerical values of the Commission's dose limits or dose constraints. Ideally, they should be the result of a generic optimisation of protection. In practice, this is unrealistically difficult and it is simpler to choose the initial values as a percentile point on the observed distribution of doses to patients. The values should be selected by professional medical bodies (in conjunction with national health and radiological protection authorities), and reviewed at intervals that represent a compromise between the necessary stability and the long-term changes in the observed dose distributions. The selected values could be specific to a country or region.

(84) In principle, it may be possible to choose a lower diagnostic reference level below which the doses would be too low to provide a sufficiently good image quality. However, such diagnostic reference levels are very difficult to set, because factors other than dose also influence image quality. Nevertheless, if the observed doses or administered activities are consistently well below the diagnostic reference level, there should be a local review of the quality of the images obtained.

10.2. Diagnostic reference levels (*Supporting Guidance 2*)

(85) More recently, in *Supporting Guidance 2* (ICRP, 2001), additional advice was provided, as noted below in paragraphs (86)–(94). ICRP (2001) also includes a survey of the various approaches that have been taken by authorised bodies, working in concert with professional medical groups, to establish diagnostic reference levels for medical imaging tasks.

(86) The objective of a diagnostic reference level is to help avoid radiation dose to the patient that does not contribute to the clinical purpose of a medical imaging task. This is accomplished by comparison between the numerical value of the diagnostic reference level (derived from relevant regional, national, or local data) and the mean or other appropriate value observed in practice for a suitable reference group of patients or a suitable reference phantom. A reference group of patients is usually defined within a certain range of physical parameters (e.g. height, weight). If an unselected sample of patients was used as a reference group, it would be unclear whether the observed value for the sample was appropriate for comparison with the diagnostic reference level. A diagnostic reference level is used for a given medical imaging task or protocol, and is not applied to individual patients.

(87) A diagnostic reference level can be used:

- to improve a regional, national, or local distribution of observed results for a general medical imaging task, by reducing the frequency of unjustified high or low values;
- to promote attainment of a narrower range of values that represent good practice for a more specific medical imaging task; or

- to promote attainment of an optimum range of values for a specified medical imaging protocol.

(88) These uses are differentiated by the degree of specification for the clinical and technical conditions selected for a given medical imaging task. Definitions and examples associated with the uses are given in *Supporting Guidance 2* (ICRP, 2001).

(89) Appropriate local review and action is taken when the value observed in practice is consistently outside the selected upper or lower level. This process helps to avoid unnecessary tissue doses being received by patients in general and, therefore, helps to avoid unnecessary risk for the associated stochastic radiation health effects.

(90) For fluoroscopically guided interventional procedures, diagnostic reference levels, in principle, could be used to promote the management of patient doses with regard to avoiding unnecessary stochastic radiation risks. However, the observed distribution of patient doses is very wide, even for a specified protocol, because the duration and complexity of the fluoroscopic exposure for each conduct of a procedure is strongly dependent on the individual clinical circumstances. A potential approach is to take into consideration not only the usual clinical and technical factors, but also the relative 'complexity' of the procedure.

(91) Diagnostic reference levels are not applicable to the management of deterministic effects (tissue reactions) (i.e. radiation-induced skin injuries) from fluoroscopically guided interventional procedures. In this case, the objective is to avoid deterministic effects (tissue reactions) in individual patients undergoing justified but long and complex procedures. The need here is to monitor, in real time, whether the threshold doses for deterministic effects (tissue reactions) are being approached or exceeded for the actual procedure as conducted on a particular patient. The relevant risk quantity is absorbed dose in the skin at the site of maximum cumulative skin dose. A helpful approach is to select values for maximum cumulative absorbed dose in the skin at which various clinical actions regarding the patient's record or care (related to potential radiation-induced skin injuries) are taken (ICRP, 2000b). Then, during actual procedures, appropriate quantities that can help to indicate the maximum cumulative absorbed dose in the skin are monitored.

(92) Diagnostic reference levels should be used to help manage the radiation dose to patients so that the dose is commensurate with the clinical purpose.

(93) The concept of a diagnostic reference level permits flexibility in the choice of quantities, numerical values, and technical or clinical specifications in order to meet the objectives relevant to the local circumstances. The guiding principles for setting a diagnostic reference level are:

- the regional, national, or local objective is clearly defined, including the degree of specification of clinical and technical conditions for the medical imaging task;
- the selected value of the diagnostic reference level is based on relevant regional, national, or local data;
- the quantity used for the diagnostic reference level can be obtained in a practical way;

- the quantity used for the diagnostic reference level is a suitable measure of the relative change in patient tissue doses and, therefore, of the relative change in patient risk for the given medical imaging task; and
- the manner in which the diagnostic reference level is to be applied in practice is clearly illustrated.

(94) Professional medical bodies (in conjunction with national health and radiological protection authorities) are encouraged to set diagnostic reference levels that best meet their specific needs and that are consistent for the regional, national, or local area to which they apply.

10.3. References

ICRP, 1991a. 1990 Recommendations of the International Commission on Radiological Protection. ICRP Publication 60. Ann. ICRP 21(1–3).

ICRP, 1996. Radiological protection and safety in medicine. ICRP Publication 73. Ann. ICRP 26(2).

ICRP, 2000b. Avoidance of radiation injuries from medical interventional procedures. ICRP Publication 85. Ann. ICRP 30(2).

ICRP, 2001. Radiation and your patient: a guide for medical practitioners. Also includes: Diagnostic reference levels in medical imaging – review and additional advice. ICRP Supporting Guidance 2. Ann. ICRP 31(4).

11. INDIVIDUAL DOSE LIMITS

(95) It is not appropriate to apply dose limits to medical exposure of patients, because such limits would often do more harm than good. Often, there are concurrent chronic, severe, or even life-threatening medical conditions that are more critical than the radiation exposure. The emphasis is then on justification of the medical procedures and on the optimisation of radiological protection.

12. PREVENTING ACCIDENTS IN RADIATION THERAPY

(96) This section discusses accident prevention in radiation therapy (equipment and procedures) (ICRP, 2000c, 2005a,b).

(97) Accident prevention in radiation therapy should be an integral part of the design of equipment and premises, and of the working procedures. A key feature of accident prevention has long been the use of multiple safeguards against the consequences of failures. This approach, called 'defence in depth', aims to prevent equipment failures and human errors, and mitigates their consequences should they happen. Some defences are provided by the design of equipment, others by the working procedures.

(98) Radiation therapy equipment should be designed to reduce operator errors by automatically rejecting demands outside the design specification. In addition, enclosures should be designed to exclude staff during exposures, without unduly isolating the patient.

(99) Radiation therapy equipment should be calibrated after installation and after any modification, and should be routinely checked by a standard test procedure that will detect significant changes in performance, as described in *Publication 86* (ICRP, 2000c).

(100) Working procedures should require key decisions, especially in radiation therapy, to be subject to independent confirmation. The patient's identity and the correct link to the prescribed treatment should be double-checked. In therapeutic nuclear medicine, dual checks should be made on the correctness of the pharmaceutical and its activity. Effective communication between all the staff involved is a vital part of the process.

(101) Radioactive sources used for therapy can cause very serious exposures if they are mislaid or misused. Brachytherapy sources should be subject to frequent and thorough accounting checks, and provision should be made for their eventual safe disposal. The possible presence of implanted sources or therapeutic activities of radiopharmaceuticals should be taken into account in the handling of deceased patients.

12.1. References

ICRP, 2000c. Prevention of accidental exposures to patients undergoing radiation therapy. ICRP Publication 86. Ann. ICRP 30(3).

ICRP, 2005a. Prevention of high-dose-rate brachytherapy accidents. ICRP Publication 97. Ann. ICRP 35(2).

ICRP, 2005b. Radiation safety aspects of brachytherapy for prostate cancer using permanently implanted sources. ICRP Publication 98. Ann. ICRP 35(3).

13. MANAGING ACCIDENTS AND INCIDENTS INVOLVING RADIOACTIVE MATERIALS

(102) This section discusses remedial actions that can be taken to reduce doses, or their consequences, resulting from an accident or from the misuse of a radioactive material. However, accidents and errors may also occur with x-ray generators and accelerators. While termination of such exposures ends the irradiation, the excess doses or their consequences may require medical treatment.

(103) In fractionated radiation therapy, an error in an early fraction can be partly corrected by adjusting further fractions. This is best thought of as part of dose planning rather than medical intervention.

(104) The misadministration of radiopharmaceuticals in diagnostic nuclear medicine does not usually cause a serious health problem, but does need to be explained fully to the patient.

(105) Several examples of remedial actions in emergency situations associated with the use of radioactive materials in medicine are as follows.

- The dose from an excessive or erroneous administration of radioiodine in therapy may be reduced by the early administration of stable iodine as potassium iodide or iodate to reduce the uptake of radioiodine by the thyroid.
- The dose from a missing brachytherapy source can be reduced by measures to locate the source and warnings to those who may be exposed.
- The dose from a major spill of radioactive materials in nuclear medicine may be reduced by the early isolation of the contaminated area and by the controlled evacuation of staff and patients.
- The doses resulting from the improper disposal and subsequent damage or mishandling of a teletherapy source may be both serious and widespread. Major countermeasures in the public domain may have to include evacuation, destruction of property, and decontamination of substantial areas. A widespread monitoring programme will be indispensable. Guidance on the levels of averted dose that would justify such actions is given in *Publication 63* (ICRP, 1993).

13.1. Reference

ICRP, 1993. Principles for intervention for protection of the public in a radiological emergency. ICRP Publication 63. Ann. ICRP 22(4).

14. EDUCATION AND TRAINING

(106) There should be radiological protection training requirements for physicians, dentists, and other health professionals who order, conduct, or assist in medical or dental procedures that utilise ionising radiation in diagnostic and interventional procedures, nuclear medicine, and radiation therapy. The final responsibility for the radiation exposure lies with the physician, who should be aware of the risks and benefits of the procedures involved.

(107) Relative to radiation use in medicine, three distinct categories of physicians can be identified:

- physicians that are trained in the ionising radiation medical specialties (e.g. radiologists, nuclear medicine physicians, radiation oncologists);
- other physicians that utilise ionising radiation modalities in their practice (e.g. cardiologists, vascular surgeons, urologists); and
- physicians that prescribe medical procedures that use ionising radiation.

(108) Education and training, appropriate to the role of each category of physician, should be given at medical schools, during the residency, and in focused specific courses. There should be an evaluation of the training, and appropriate recognition that the individual has completed the training successfully. In addition, there should be corresponding radiological protection training requirements for other clinical personnel that participate in the conduct of procedures utilising ionising radiation, or in the care of patients undergoing diagnosis or treatments with ionising radiation.

15. INSTITUTIONAL ARRANGEMENTS

(109) In particular, it is important to clarify the separate responsibilities of the referring physicians who request radiological procedures, the radiologists and other practitioners who undertake the procedures, and the administrators who provide the resources. The role of medical physicists also should be incorporated.

(110) One important need is to provide adequate resources for education and training in radiological protection for future professional and technical staff who request or partake in radiological practices in medicine. The training programme should include initial training for all incoming staff, regular updating and retraining, and certification of the training.

(111) Quality assurance programmes are essential for maintaining the intended standards in all the functions of the undertaking. Their scope should specifically include radiological protection and safety. Quality assurance and audit programmes (including patient dose or administered activity assessments) and inspections by competent authorities should ensure that medical exposure is delivered under good radiation protection conditions. These programmes should include acceptance testing carried out before the first use of the equipment for clinical purposes, and thereafter performance testing on a regular basis and after any major maintenance procedure. Specific quality assurance programmes should be adapted to the new imaging technologies (e.g. digital) and radiation therapy procedures.

(112) Any system of verification includes record keeping. The requirements for recording occupational exposures will usually be determined by the regulatory authorities. When procedure-related dose quantities are measured for diagnostic or interventional procedures, records should be kept of any comparisons with diagnostic reference levels. In radiation therapy, the data from dose planning, administered activity (in nuclear medicine), and, for radiation therapy patients, the activity at the time of discharge should be included in the patients' records.

16. PRACTICAL METHODS OF PROTECTION OTHER THAN FOR PATIENTS

16.1. Occupational exposure

(113) The principles for the protection of workers from ionising radiation, including in medicine, are discussed fully in *Publication 75* (ICRP, 1997). These principles apply to staff in x-ray, nuclear medicine, and radiation therapy facilities.

(114) The control of occupational exposure can be simplified and made more effective by the designation of workplaces into two types: controlled areas and supervised areas. In a controlled area, normal working conditions, including the possible occurrence of minor mishaps, require workers to follow well-established procedures and practices aimed specifically at controlling radiation exposures. A supervised area is one in which the working conditions are kept under review, but special procedures are not normally needed. The definitions are best based on operational experience and judgement. In areas where there is no problem of contamination by unsealed radioactive materials, designated areas may sometimes be defined in terms of the dose rate at the boundary.

(115) Individual monitoring for external radiation is fairly simple and does not require a heavy commitment of resources. In medicine, it should be used for all those who work in controlled areas.

(116) In several areas of medicine, the control of occupational exposure is of particular importance. One of these is the nursing of brachytherapy patients when the sources have been implanted, rather than inserted by after-loading techniques. A second is palpation of patients during procedures utilising fluoroscopy. A third is in fluoroscopically guided interventional procedures, as in heart catheterisation. A fourth is radiopharmaceutical preparation by staff in nuclear medicine. In all these procedures, careful shielding and time limits are needed. Individual monitoring with careful scrutiny of the results is also important. In brachytherapy, frequent and careful accounting of sources is essential.

(117) The system for protecting staff from the source (e.g. shielding) should be designed to minimise any sense of isolation experienced by the patient. This is particularly relevant in nuclear medicine and brachytherapy, where the source is within the patient.

(118) The Commission recommends that the working conditions of a pregnant worker, after the declaration of pregnancy, should be such as to make it unlikely that the additional equivalent dose to the embryo/fetus will exceed approximately 1 mSv during the remainder of the pregnancy. In the interpretation of this recommendation, it is important not to create unnecessary discrimination against pregnant women. The part of a pregnancy prior to declaration of the pregnancy is covered by the normal protection of workers, which is essentially the same for females and males.

16.2. Public exposure

(119) Public access to hospitals and radiology rooms is restricted, but it is more open than is common in industrial and research laboratory operations. There are no radiological protection grounds for imposing restrictions on public access to non-designated areas. Due to the limited duration of public access, an access policy can be adopted for supervised areas if this is of benefit to patients or visitors and there are appropriate radiological protection safeguards. Public access to controlled areas with high-activity sources (e.g. brachytherapy and other therapy sources) should be limited to patients' visitors, who should be advised of any restrictions on their behaviour.

16.3. Exposure of volunteers in biomedical research

(120) The use of volunteers in biomedical research makes a substantial contribution to medicine and human radiobiology. Some research studies are of direct value in the investigation of disease, and others provide information on the metabolism of pharmaceuticals and radionuclides that may be absorbed from contamination of the workplace or the environment. Not all of these studies take place in medical institutions, but the Commission includes the exposure of all these volunteers under the category of medical exposure.

(121) The ethical and procedural aspects of the use of volunteers in biomedical research have been addressed by the Commission in *Publication 62* (ICRP, 1991b). The key aspects include the need to guarantee a free and informed choice by the volunteers, the adoption of dose constraints linked to the societal worth of the studies, and the use of an ethics committee that can influence the design and conduct of the studies. The use of children and the mentally ill or defective in biomedical research is also addressed in *Publication 62* (ICRP, 1991b). It is important that the ethics committee should have easy access to radiological protection advice.

(122) In many countries, radiation exposure of pregnant females in biomedical research is not specifically prohibited. However, their involvement in such research is very rare and should be discouraged unless pregnancy is an integral part of the research. In order to protect the embryo/fetus, strict controls should be placed on the use of radiation in these cases.

16.4. Exposure of comforters and carers of patients

(123) Friends and relations helping in the support and comfort of patients are also volunteers, but there is a direct benefit both to the patient and those who care for them. Their exposures are defined as medical exposures, but dose constraints should be established for use in defining the protection policy for visitors to patients, and families at home when nuclear medicine patients are discharged from hospital. Such groups may include children. The Commission has not previously recommended values for such constraints, but a value of 5 mSv per episode for an adult (i.e. for the duration of a given release of a patient after therapy) is reasonable. The constraint

needs to be used flexibly. For example, higher doses may well be appropriate for the parents of very sick children. Young children, infants, and visitors not engaged in direct comforting or care should be treated as members of the public (subject to the public dose limit of 1 mSv/year). The topic of release of patients after therapy with unsealed radionuclides is covered in further detail in Section A.7.

16.5. References

ICRP, 1991b. Radiological protection in biomedical research. ICRP Publication 62. Ann. ICRP 22(3).

ICRP, 1997. General principles for the radiation protection of workers. ICRP Publication 75. Ann. ICRP 27(1).

ANNEX A. FOCUSED EVALUATIONS OF RADIOLOGICAL PROTECTION IN MEDICINE

(A1) Committee 3 has produced a number of documents that provide detailed advice related to radiological protection and safety in the medical applications of ionising radiation. Each document focuses on a particular radiation source as applied in a given medical discipline or to a given type of patient. Each document is a compendium of the application of the extant Commission recommendations, as applicable to medical radiation. In brief, the following observations appear to be predominant with regard to radiological protection and safety in medicine.

- Communications must be directed to the relevant medical practitioners, in a format in which they are conversant, and channelled to them by an appropriate authoritative or professional body.
- In diagnostic and interventional procedures, management of the patient dose commensurate with the medical task is the appropriate mechanism to avoid unproductive radiation exposure. Equipment features that allow this to be accomplished, and diagnostic reference levels derived at the appropriate national, regional, or local level are likely to be the most effective approaches.
- In radiation therapy, the avoidance of accidents is the predominant issue. A review of such accidents and advice for accident prevention is found in *Publication 86* (for external beam and solid brachytherapy sources) (ICRP, 2000c), *Publication 97* [additional advice for high-dose-rate (HDR) brachytherapy sources] (ICRP, 2005a), and *Publication 98* (additional advice for permanently implanted sources used in brachytherapy for prostate cancer) (ICRP, 2005b).

(A2) Brief synopses of these focused publications are provided below in the order in which the documents were published. Each illustrates the aspects of the Commission's radiological protection framework that are most relevant.

A.1. Pregnancy and medical radiation (*Publication 84*)

(A3) Thousands of pregnant patients and radiation workers are exposed to ionising radiation each year. Lack of knowledge is responsible for great anxiety and probably unnecessary termination of pregnancies. For many patients, the exposure is appropriate, while for other patients, the exposure may be inappropriate, placing the embryo/fetus at increased risk.

(A4) Before any exposure using ionising radiation, it is important to determine whether a female is, or could be, pregnant. Medical exposures during pregnancy require specific consideration due to the radiation sensitivity of the developing embryo/fetus. The manner in which an examination is performed depends on whether the embryo/fetus will be in the direct beam and whether the procedure requires a relatively high dose.

(A5) Prenatal doses from most correctly performed diagnostic procedures present no measurably increased risk of prenatal death, developmental damage including malformation, or impairment of mental development over the background incidence of these entities. Higher doses, such as those involved in therapeutic procedures, have the potential to result in developmental harm.

(A6) The pregnant patient or worker has a right to know the magnitude and type of potential radiation effects that may result from in-utero exposure. Almost always, if a diagnostic radiology examination is medically indicated, the risk to the mother of not performing the procedure is greater than the risk of potential harm to the embryo/fetus. Most nuclear medicine procedures do not result in high doses to the embryo/fetus. However, some radiopharmaceuticals that are used in nuclear medicine (e.g. radioiodides) can pose increased risks to the embryo/fetus.

(A7) It is essential to ascertain whether a female patient is pregnant prior to radiation therapy. In pregnant patients, cancers that are remote from the pelvis can usually be treated with radiation therapy. However, this requires careful planning. Cancers in the pelvis cannot be treated adequately during pregnancy without severe or lethal consequences for the embryo/fetus.

(A8) The basis for the control of occupational exposure of women who are not pregnant is the same as that for men. However, if a woman declares to her employer that she is pregnant, additional controls have to be considered in order to attain a level of protection for the embryo/fetus broadly similar to that provided for members of the public.

(A9) In many countries, radiation exposure of pregnant females in biomedical research is not specifically prohibited. However, their involvement in such research is very rare and should be discouraged unless pregnancy is an integral part of the research. In order to protect the embryo/fetus, strict controls should be placed on the use of radiation in these cases.

(A10) Termination of pregnancy is an individual decision affected by many factors. Absorbed doses below 100 mGy to the developing organism should not be considered a reason for terminating a pregnancy. At doses to the embryo/fetus above this level, informed decisions should be made based upon individual circumstances, including the magnitude of the estimated dose to the embryo/fetus, and the consequent risks of harm to the developing embryo/fetus and risks of cancer in later life.

A.2. Interventional procedures (fluoroscopically guided) (*Publication 85*)

(A11) Fluoroscopically guided interventional procedures are being used by an increasing number of clinicians who are not adequately trained in radiation safety or radiobiology. Many of these interventionists are not aware of the potential for injury from these procedures or the simple methods for decreasing their incidence. Many patients are not being counselled on the radiation risks, nor followed-up when radiation doses from difficult procedures may lead to injury. Some patients are suffering radiation-induced skin injuries, and younger patients may face an increased risk of future cancer. Interventionists are having their practice limited or suffering injury, and are exposing their staff to high doses.

(A12) In some of these interventional procedures, skin doses to patients approach those experienced in radiation therapy fractions in the treatment of cancer. Radiation-induced skin injuries are occurring in patients due to the use of inappropriate equipment and, more often, poor operational technique. Injuries to physicians and staff performing these interventional procedures have also been observed. Acute radiation doses (to patients) may cause erythaema at 2 Gy, cataract at 2 Gy, permanent epilation at 7 Gy, and delayed skin necrosis at 12 Gy. Protracted (occupational) exposures to the eye may cause cataracts at 4 Gy if the dose is received in less than 3 months, and at 5.5 Gy if received over a period exceeding 3 months. However, new data on the radiosensitivity of the eye with regard to visual impairment are expected, and the Commission will consider these data when they become available.

(A13) Practical actions to control dose to the patient and to the staff are available. The absorbed dose to the patient in the area of skin that receives the maximum dose is of major concern. Each local clinical protocol should include, for each type of fluoroscopically guided interventional procedure, a statement on the cumulative skin doses and skin sites associated with the various parts of the procedure. Interventionists should be trained to use information on skin dose and on practical techniques to control dose. Maximum cumulative absorbed doses that appear to approach or exceed 1 Gy (for procedures that may be repeated) or 3 Gy (for any procedure) should be recorded in the patient's record, and there should be a follow-up procedure for such cases. Patients should be counselled if there is a significant risk of radiation-induced injury, and the patient's personal physician should be informed of the possibility of radiation effects. Training in radiological protection for patients and staff should be an integral part of the education for those using these interventional procedures. All interventionists should audit and review the outcomes of their procedures for radiation injury. Risks and benefits, including radiation risks, should be taken into account when new fluoroscopically guided interventional techniques are introduced.

A.3. Accidental exposures in radiation therapy (*Publication 86*)

(A14) From the viewpoint of radiation safety, radiation therapy is a very special application of radiation because:

- human beings are directly placed in a very intense radiation beam (external beam therapy), or radiation sources are placed in direct contact with tissue (brachytherapy), to deliver very high doses (20–80 Gy) intentionally; and
- overdosage as well as underdosage may have severe consequences.

(A15) *Publication 86* intends to assist in the prevention of accidental exposures involving patients undergoing treatment from external beam or solid brachytherapy sources. It does not deal directly with radiation therapy involving unsealed sources. The document is addressed to a diverse audience of professionals directly involved in radiation therapy procedures, hospital administrators, and health and regulatory authorities. The approach adopted is to describe illustrative severe accidents, discuss the causes of these events and contributory factors, summarise the sometimes devastating consequences of these events, and provide recommendations on the prevention of such events. The measures discussed include institutional arrangements, staff training, quality assurance programmes, adequate supervision, clear definition of responsibilities, and prompt reporting.

(A16) In many of the accidental exposures described in this report, a single cause cannot be identified. Usually, there was a combination of factors contributing to the accident (e.g. deficient staff training, lack of independent checks, lack of quality control procedures, and absence of overall supervision). Such combinations often point to an overall deficiency in management, allowing patient treatment in the absence of a comprehensive quality assurance programme. Factors common to many accidents are identified and discussed in detail, and explicit recommendations on measures to prevent radiation therapy accidents are given with respect to regulations, education, and quality assurance.

(A17) Doses received during radiation therapy are on the upper edge of tolerable doses to normal tissues. As a result, accidental overdosages have often had devastating and sometimes fatal consequences. Accidental exposures involving an overdosage of 10% or more should be detectable by a well-trained clinician, based upon an unusually high incidence of adverse patient reactions. Underdosage accidents are difficult to detect clinically and may only be manifest as poor tumour control.

(A18) The frequency of radiation therapy is increasing worldwide, and accidents may be expected to increase in frequency if measures for prevention are not taken. While a number of serious and fatal radiation therapy accidents are reported, it is likely that many more have occurred but were either not recognised or reported to regulatory authorities or published in the literature.

(A19) The complex equipment and techniques used in radiation therapy mandate that for accident prevention, there must be sound and risk-informed regulations, managerial commitment at the hospital level, an adequate number of trained staff, adequate resources, a functional implemented quality assurance programme, good communication, and continuing education.

(A20) There is a danger in not fully appreciating that modern equipment and new technologies require more quality assurance and highly qualified maintenance. Persons in charge of radiation therapy facilities should ensure that there is proper commissioning of new equipment and proper decommissioning of old equipment and sources.

A.4. Computed tomography (*Publication 87*)

(A21) CT examinations can involve relatively high doses to patients. The cumulative absorbed doses to tissues from multiple CT procedures (10–100 mGy/procedure) can often approach or exceed the levels known, from epidemiological studies, to increase the probability of cancer. The frequency of CT examinations is increasing worldwide and the types of examination using CT are also becoming more numerous. However, in contrast to the common trend in diagnostic radiology, the rapid developments in CT have not led in general to a reduction of patient doses for a given type of application.

(A22) Therefore, management of patient dose is crucial. The referring physician should evaluate whether the result of each examination will affect the clinical management of the patient. The radiologist should concur that the procedure is justified. The radiologist and CT system operator should be aware of the possibilities to reduce patient doses by adapting technical parameters to each patient and the examination at hand, with special attention being paid to paediatric and young patients. A reduction in patient dose of more than 50% is possible by an appropriate choice of technical parameters, attention to quality control, and the application of diagnostic reference levels in co-operation with a medical physicist. Further improvements in CT equipment could help the operator to reduce unnecessary patient doses substantially. The most important of these features will be anatomically based on-line adjustment of exposure factors, and new image reconstruction approaches associated with multi-slice CT.

A.5. Guide for general practitioners (*Supporting Guidance 2*)

(A23) This didactic text is devoted to the protection of patients against unnecessary exposure to ionising radiation. It is organised in a questions-and-answers format.

(A24) There are obvious benefits to health from medical uses of radiation, in x-ray diagnostic examinations, interventional procedures, nuclear medicine, and radiation therapy. However, there are well-established risks from high doses of radiation (radiation therapy, interventional procedures), particularly if improperly applied, and possible deleterious effects from small radiation doses (such as those used in diagnostics). Appropriate use of large doses in radiation therapy prevents serious harm, but even low doses carry a risk that cannot be eliminated entirely. Therefore, diagnostic use of radiation requires such methodology that would secure high diagnostic gains while minimising the possible harm.

(A25) The text provides ample information on opportunities to minimise doses, and therefore the risk from diagnostic uses of radiation. This objective may be reached by avoiding unnecessary (unjustified) examinations, and by optimising the procedures applied both from the standpoint of diagnostic quality and in terms of reduction of the excessive doses to patients.

(A26) Optimisation of patient protection in radiation therapy must depend on maintaining sufficiently high doses to irradiated tumours, securing a high cure rate, while protecting the healthy tissues as much as possible.

(A27) Problems related to special protection of the embryo/fetus in the course of diagnostic and therapeutic uses of radiation are presented, and practical solutions are recommended.

A.6. Digital radiology (*Publication 93*)

(A28) Digital techniques have the potential to improve the practice of radiology but they also risk the overuse of radiation. The main advantages of digital imaging (i.e. wide dynamic range, post processing, multiple viewing options, and electronic transfer and archiving possibilities) are clear, but overexposures can occur without an adverse impact on image quality. In conventional radiography, excessive exposure produces a 'black' film. In digital systems, good images are obtained for a large range of doses. It is very easy to obtain (and delete) images with digital fluoroscopy systems, and there may be a tendency to obtain more images than necessary.

(A29) In digital radiology, higher patient dose usually means improved image quality, so a tendency to use higher patient doses than necessary could occur. Different medical imaging tasks require different levels of image quality, and doses that have no additional benefit for the clinical purpose should be avoided.

(A30) Image quality can be compromised by inappropriate levels of data compression and/or post-processing techniques. All these new challenges should be part of the optimisation of protection process, and should be included in clinical and technical protocols.

(A31) Local diagnostic reference levels should be re-evaluated for digital imaging, and patient dose parameters should be displayed at the operator console. Frequent patient dose audits should occur when digital techniques are introduced. Training in the management of image quality and patient dose in digital radiology is necessary. Digital radiology will involve new regulations and invoke new challenges for practitioners. As digital images are easier to obtain and transmit, the justification criteria should be reinforced.

(A32) Commissioning of digital systems should involve clinical specialists, medical physicists, and radiographers to ensure that imaging capability and radiation dose management are integrated. Quality control requires new procedures and protocols (visualisation, transmission, and archiving of the images).

(A33) Industry should promote tools to inform radiologists, radiographers, and medical physicists about the exposure parameters and the resultant patient doses associated with digital systems. The exposure parameters and the resultant patient doses should be standardised, displayed, and recorded.

A.7. Unsealed radionuclides (release after therapy) (*Publication 94*)

(A34) After some therapeutic nuclear medicine procedures with unsealed radionuclides, precautions may be needed to limit doses to other people, but this is rarely the case after diagnostic procedures. ^{131}I results in the largest dose to medical staff, the public, caregivers, and relatives. Other radionuclides used in therapy are usually simple beta emitters (e.g. ^{32}P, ^{89}Sr, and ^{90}Y) that pose much less risk. Dose limits apply to exposure of the public and medical staff resulting from medical exposure of patients.

(A35) Previously, the Commission recommended that a source-related dose constraint of a few mSv per episode applies to relatives, visitors, and caregivers at home, rather than a dose limit (*Publication 73*) (ICRP, 1996). A dose constraint of 5 mSv/episode (i.e. for the duration of a given release of a patient after therapy) is likely to be reasonable (see Section 16.4).

(A36) *Publication 94* (ICRP, 2004) recommends that young children and infants, as well as visitors not engaged in direct care or comforting, should be treated as members of the public (i.e. be subject to the public dose limit of 1 mSv/year).

(A37) The modes of exposure to other people are: external exposure; internal exposure due to contamination; and environmental pathways. Dose to adults from patients is mainly due to external exposure. Contamination of infants and children with saliva from a patient could result in significant doses to the child's thyroid. It is important to avoid contamination of children and pregnant women. After radioiodine therapy, mothers must cease breastfeeding immediately. Many types of therapy with unsealed radionuclides are contraindicated in pregnant females. Women should not become pregnant for some time after radionuclide therapy (e.g. 6 months for radioiodine, the most common radionuclide used). Various shorter or longer times for other radionuclides are given in *Publication 94* (ICRP, 2004).

(A38) ^{99m}Tc dominates discharges to the environment from excreta of nuclear medicine patients, but its short half-life limits its importance. The second largest discharges, ^{131}I, can be detected in the environment after medical uses but with no measurable environmental impact. Storing patients' urine after radionuclide therapy appears to have minimal benefit. Radionuclides released into modern sewage systems are likely to result in doses to sewer workers and the public that are well below public dose limits.

(A39) The decision to hospitalise or release a patient should be determined on an individual basis. In addition to residual activity in the patient, the decision should take many other factors into account. Hospitalisation will reduce exposure to the public and relatives, but will increase exposure to hospital staff. Hospitalisation often involves a significant psychological burden as well as monetary and other costs that should be analysed and justified. Patients travelling after radioiodine therapy rarely present a hazard to other passengers if travel times are limited to a few hours.

(A40) Environmental or other radiation-detection devices are able to detect patients who have had radioiodine therapy for several weeks after treatment. Personnel operating such detectors should be specifically trained to identify and deal with nuclear medicine patients. Records of the specifics of therapy with unsealed radionuclides should be maintained at the hospital and given to the patient along with written precautionary instructions. In the case of death of a patient who has had therapy with unsealed radionuclides in the last few months, special precautions may be required.

A.8. High-dose-rate brachytherapy (accidents) (*Publication 97*)

(A41) HDR brachytherapy is a rapidly growing technique that has been replacing low-dose-rate (LDR) procedures over the last few years in both industrialised and developing countries. It is estimated that approximately 500,000 procedures (administrations of treatment) are performed by HDR units annually. LDR equipment has been discontinued by many manufacturers, leaving HDR brachytherapy as the major alternative.

(A42) HDR brachytherapy techniques deliver a very high dose, of the order of 1.6–5.0 Gy/min, so mistakes can lead to under- or overdosage with the potential for clinical adverse effects. More than 500 HDR accidents (including one death) have been reported along the entire chain of procedures from source packing to delivery of dose. Human error has been the prime cause of radiation events. The Commission concludes that many accidents could have been prevented if staff had had functional monitoring equipment and paid attention to the results.

(A43) Since ^{192}Ir has a relatively short half-life, the HDR sources need to be replaced approximately every 4 months. Over 10,000 HDR sources are transported annually, with the resultant potential for accidents; therefore, appropriate procedures and regulations must be observed.

(A44) A number of specific recommendations on procedures and equipment are given in this report. The need for an emergency plan and for practicing emergency procedures is stressed. The possibility of loss or theft of sources must be kept in mind.

(A45) A collaborating team of specifically trained personnel following quality assurance procedures is necessary to prevent accidents. Maintenance is an indispensable component of quality assurance; external audits of procedures reinforce good and safe practice, and identify potential causes of accidents. Quality assurance should include peer review of cases. Accidents and incidents should be reported and the lessons learned should be shared with other users to prevent similar mistakes.

A.9. Brachytherapy for prostate cancer with permanent sources (radiation safety) (*Publication 98*)

(A46) The use of permanent radioactive implants (^{125}I or ^{103}Pd seeds) to treat selected localised prostate cancer patients has been increasing rapidly all over the world for the last 15 years. It is estimated that more than 50,000 patients receive this treatment annually worldwide, and this number is anticipated to increase in the near future.

(A47) Although no accidents or adverse effects involving medical staff and members of the patient's family have been reported to date, this brachytherapy technique raises a number of radiation safety issues.

(A48) All data concerning the dose received by people approaching patients after implantation have been reviewed. Those doses have either been measured directly or calculated. The available data show that, in the vast majority of cases, the dose to comforters and carers is well below 1 mSv/year. Only the (rare) case where the patient's partner is pregnant at the time of implantation may need specific precautions.

(A49) Expulsion of sources through urine, semen, or the gastrointestinal tract is rare. Specific recommendations should be given to patients to allow them to deal with this event adequately. Of note, due to the low activity of an isolated seed and its low photon energy, no incident or accident linked to seed loss has ever been recorded.

(A50) Cremation of a corpse (common in some countries) in the first few months after implantation raises several issues related to: the activity that remains in the patient's ashes; and the airborne dose, potentially inhaled by crematorium staff or members of the public. Review of available data shows that cremation can be allowed if 12 months have elapsed since implantation with ^{125}I (3 months for

^{103}Pd). If the patient dies before this delay has elapsed, specific measures must be undertaken.

(A51) Specific recommendations have to be given to the patient to warn his surgeon in case of subsequent pelvic or abdominal surgery. A 'wallet card' with all relevant information about the implant is useful.

(A52) In most cases, brachytherapy does make the patient infertile. However, although the therapy-related modifications of the semen reduce fertility, patients must be aware of the possibility of fathering children after such a permanent implantation, with a limited risk of genetic effects for the child.

(A53) Patients with permanent implants must be aware of the possibility of triggering certain types of security radiation monitors. The 'wallet card' including the main information about the implant (see above) may prove to be helpful in such a case.

(A54) Considering the available experience after brachytherapy and external irradiation of prostate cancer, the risk of radio-induced secondary tumours appears to be extremely low. The demonstrated benefit of brachytherapy clearly outweighs, by far, the very limited (mainly theoretical) increase in the radiation-induced cancer risk.

A.10. Multi-detector computed tomography (*Publication 102*)

(A55) Modern CT scanners employ multiple rows of detector arrays allowing rapid scanning and wider scan coverage. All new CT systems have multiple detectors (MDCT) with a single or dual x-ray source, and a number of new dose-reduction tools have become available commercially.

(A56) There are a number of aspects specific to MDCT that systematically increase or decrease patient dose compared with single-detector row CT scanners (SDCT). Initial reports indicated increased patient doses for MDCT relative to SDCT, but more recent reports have shown comparable or decreased patient doses. If the user selects settings for MDCT identical to those used in SDCT, there can be an increase in patient dose. Settings appropriate to a specific scanner model must be determined.

(A57) There is potential for dose reduction with MDCT systems, but the actual dose reduction achieved depends upon how the system is used. It is important that radiologists, cardiologists, medical physicists, and CT system operators understand the relationship between patient dose and image quality, and are aware that image quality in CT is often higher than that needed for diagnostic confidence. Images of the highest quality are not essential for all diagnostic tasks, but rather the level of quality (e.g. low noise, medium, or low dose) is dependent on the diagnostic task.

(A58) There is increasing awareness of how adapting exposure factors can contribute to the management of patient dose. However, the rate at which technology is changing requires continued attention to management of patient dose.

(A59) Automatic exposure control systems enable scan protocols to be applied using measures related to image quality. If the image quality is appropriately specified by the user and suited to the clinical task, there is a reduction in patient dose for all but the obese patient. In obese patients, the dose is increased to improve the image quality. Automatic exposure control does not totally free the operator from selection of scan parameters, and awareness of individual systems is important.

(A60) The selection of image quality parameters in automatic exposure control systems is not straightforward. There is a lack of consensus on how image quality is to be specified, and there are significant differences in the way that different companies achieve exposure control. Operator knowledge of the system is important.

(A61) Justification of CT use is a shared responsibility between requesting clinicians and radiologists. It includes justification of the CT study for a given indication, and classification of the clinical indications into those requiring standard-dose CT and those requiring low-dose CT. Scanning parameters should be based on study indication, patient size, and body region being scanned so that patient dose can be managed based on these parameters. Guidelines (selection criteria for CT examinations) are necessary so that inappropriate studies can be avoided. In addition, alternate non-radiation imaging techniques should be considered, when appropriate.

(A62) With the emergence of cardiac MDCT applications, many cardiologists have become users of MDCT scanners. The Commission recommends appropriate training in radiation protection for cardiologists. Training of requesting physicians and CT staff can help in the management of scan indications, protocols, and patient dose.

A.11. References

ICRP, 1996. Radiological protection and safety in medicine. ICRP Publication 73. Ann. ICRP 26(2).

ICRP, 2000c. Prevention of accidental exposures to patients undergoing radiation therapy. ICRP Publication 86. Ann. ICRP 30(3).

ICRP, 2004. Release of patients after therapy with unsealed radionuclides. ICRP Publication 94. Ann. ICRP 34(2).

ICRP, 2005a. Prevention of high-dose-rate brachytherapy accidents. ICRP Publication 97. Ann. ICRP 35(2).

ICRP, 2005b. Radiation safety aspects of brachytherapy for prostate cancer using permanently implanted sources. ICRP Publication 98. Ann. ICRP 35(3).

ALL REFERENCES

ICRP, 1991a. 1990 Recommendations of the International Commission on Radiological Protection. ICRP Publication 60. Ann. ICRP 21(1–3).

ICRP, 1991b. Radiological protection in biomedical research. ICRP Publication 62. Ann. ICRP 22(3).

ICRP, 1993. Principles for intervention for protection of the public in a radiological emergency. ICRP Publication 63. Ann. ICRP 22(4).

ICRP, 1996. Radiological protection and safety in medicine. ICRP Publication 73. Ann. ICRP 26(2).

ICRP, 1997. General principles for the radiation protection of workers. ICRP Publication 75. Ann. ICRP 27(1).

ICRP, 1999a. Genetic susceptibility to cancer. ICRP Publication 79. Ann. ICRP 28(1/2).

ICRP, 1999b. Radiation dose to patients from radiopharmaceuticals. Addendum to ICRP Publication 53. Also includes Addendum 1 to ICRP Publication 72. ICRP Publication 80. Ann. ICRP 28(3).

ICRP, 2000a. Pregnancy and medical radiation. ICRP Publication 84. Ann. ICRP 30(1).

ICRP, 2000b. Avoidance of radiation injuries from medical interventional procedures. ICRP Publication 85. Ann. ICRP 30(2).

ICRP, 2000c. Prevention of accidental exposures to patients undergoing radiation therapy. ICRP Publication 86. Ann. ICRP 30(3).

ICRP, 2000d. Managing patient dose in computed tomography. ICRP Publication 87. Ann. ICRP 30(4).

ICRP, 2001. Radiation and your patient: a guide for medical practitioners. Also includes: Diagnostic reference levels in medical imaging – review and additional advice. ICRP Supporting Guidance 2. Ann. ICRP 31(4).

ICRP, 2003a. Managing patient dose in digital radiology. ICRP Publication 93. Ann. ICRP 34(1).

ICRP, 2003b. Biological effects after prenatal irradiation (embryo and fetus). ICRP Publication 90. Ann. ICRP 33(1/2).

ICRP, 2003c. Relative biological effectiveness (RBE), quality factor (Q), and radiation weighting factor (w_R). ICRP Publication 92. Ann. ICRP 33(4).

ICRP, 2004. Release of patients after therapy with unsealed radionuclides. ICRP Publication 94. Ann. ICRP 34(2).

ICRP, 2005a. Prevention of high-dose-rate brachytherapy accidents. ICRP Publication 97. Ann. ICRP 35(2).

ICRP, 2005b. Radiation safety aspects of brachytherapy for prostate cancer using permanently implanted sources. ICRP Publication 98. Ann. ICRP 35(3).

ICRP, 2005c. Low-dose extrapolation of radiation-related cancer risk. ICRP Publication 99. Ann. ICRP 35(4).

ICRP, 2007a. Biological and epidemiological information on health risks attributable to ionising radiation: a summary of judgements for the purposes of radiological protection of humans. Annex A to 2007 Recommendations.

ICRP, 2007b. Quantities used in radiological protection. Annex B to 2007 Recommendations.

ICRP, 2007c. Managing patient dose in multi-detector computed tomography. ICRP Publication 102. Ann. ICRP 37(1).

ICRP, 2007d. The 2007 Recommendations of the International Commission on Radiological Protection. ICRP Publication 103. Ann. ICRP 37(2–4).

ICRU, 2005. Patient dosimetry for x rays used in medical imaging. ICRU Report 74. J. ICRU 5(2).

Mettler, F.A., Thomadsen, B.R., Bhargavan, M., et al., 2008. Medical radiation exposure in the U.S. 2006: Preliminary results. 43rd Annual Meeting of the National Council on Radiation Protection and Measurements: Advances in Radiation Protection in Medicine. Health Phys. 95, in press.

NAS/NRC, 2006. Health Risks from Exposure to Low Levels of Ionising Radiation: BEIR VII Phase 2. Board on Radiation Effects Research. National Research Council of the National Academies, Washington, D.C.

UNSCEAR, 2000. Sources and Effects of Ionising Radiation. United Nations Scientific Committee on the Effects of Atomic Radiation Report to the General Assembly with Scientific Annexes, United Nations, New York, NY.